Building Stone Walls

Excerpted from *Stonework,* by Charles McRaven

CONTENTS

Why Build Stone Walls?..2

Getting to Know Stone 3

Finding a Source for Good Stone 5

Tools and Techniques for Handling Stone 8

Staying Safe..11

Cutting and Shaping Stone 12

Building a Drystone Wall...14

Building a Mortared Wall..20

Why Build Stone Walls?

Why stone? Well, a better question might be, why not? For building or landscaping, you simply can't do better. Stone is weatherproof, ratproof, insectproof, and long lived. Stone is quietly elegant and looks expensive; whether you use it in rustic or formal designs, it signifies good taste.

In this age of disposables and throwaways, stone is also psychologically appealing; it represents strength and stability. Of course, it's easier to work in wood, plastic, glass, steel, and even brick and cinder block, but once stone is in place it becomes a lifelong element of the landscape; it *belongs*. Building with stone is a tribute to permanence.

Stone has a negative aspect, however: It's *heavy*. Stone requires cement or gravel footings that are strong, deep, and wide, and sometimes you'll need elaborate lifting devices to get stones up high enough for placement on walls. Being heavy, stone also is frequently dangerous to handle. Just as gravity and friction keep it in place, they can make positioning stone laborious and hazardous.

In addition, when compared with wood or cement, stone is difficult to shape, and setting stone is unforgiving work that requires great patience. With a helper, a good mason can lay about 20 square feet (1.8 sq m) a day. Some do 30 (2.7 sq m) or more, but good, tight, artistic work takes time — time to select, fit, reject, shape, and mortar each stone.

Stonework is not for everyone. But everyone can learn from handling stone, enjoying the discipline, the craft, and the satisfaction that comes from artfully building a wall that will last for centuries.

Freestanding stone walls are often used to mark a boundary or to act as an entryway to a private space, such as a garden.

Getting to Know Stone

The stone you choose to work with should match the environment in which you are placing it. Too often, people use cut stone or freshly quarried stone, or they choose a nonnative stone because they have seen it somewhere or picked it out of a magazine. Your stone should match native stone; if you can't find any native stone to work with, look for some that matches it as closely as possible. Weathered field-stone is the most agreeable to see, because that's what you see in nature. You want character: A mossy, irregular piece of granite, a lichened sandstone, or an eroded limestone will appear to have been where it is for much of its millions of years of age. Even in formal gardens, the man-made symmetry must give the impression of age. Excessive shaping and smoothness, fresh cuts, geometric cuteness, wide mortar joints — all destroy this feeling.

Sandstone and Quartzite

Sandstones and quartzites are the most versatile building stones. They range from coarse, soft, crumbly rocks to dense, fine-grained creek quartzites so hard that they ring when struck.

Sandstone is a good stone to learn on because it cuts well, occurs in layers, and is porous enough to age quickly after shaping. It comes in as many colors as sand itself — grays, browns, whites, roses, and blues (the most common, though, are grays and browns) — and is composed of fine sand particles fused together under great pressure. Many sandstones have a definite grain along them that can be split easily. Therefore, sandstone is best laid flat, the way it was formed. Set on edge, it may weather in such a way that the layers separate. The sandstone you may have access to could be soft or hard, weak or strong. In mortared work, the stone should be at least as hard as the mortar.

Sandstone usually splits into even thicknesses in nature, so the critical top and bottom surfaces are already formed. If any shaping is necessary, it may be nothing more than a bit of nudging on the face of each stone to give an acceptable appearance. When making your selection, of course, try to find stones that are already well shaped, thereby keeping any necessary shaping to a minimum. Any sandstone can be worked to the shape you desire, but there's a logical limit. If you spend all your available time shaping, then efficiency plummets.

Limestone

Limestone has always been a favorite stone for builders. Dense but not hard, it can be worked to almost any shape. Before concrete blocks were invented (around 1900), limestone was the accepted standard for commercial stonework.

Newly cut limestone has a slick surface that is unattractive; weathered, top-of-the-ground limestone, on the other hand, is often rough, fissured, pockmarked, and interesting. (In addition, limestone often houses fossil seashells, trilobites, and other traces of prehistoric life, which can add another interesting dimension to your wall.) If you have access only to fresh-cut limestone, however, take heart: Aging is a long process, but the newly cut faces will lose their fresh-cut look in five years or so.

Granite

Granites are generally rough-textured stones that are not naturally layered. When weathered, their exterior provides a welcoming environment for lichens and mosses. Strong and hard granites vary in color. Along the East Coast, the familiar light gray granite is plentiful. Formed principally of feldspar and quartz, it's a favored landscaping stone. There are also dark blue, dark gray, greenish, and even pink granites.

If you use granite, try to find stones naturally endowed with the desired shapes, as they can be hard to shape. Granite can often be recycled from foundations, chimneys, and basements of abandoned buildings.

Stones to Avoid

Shale, slate, and other soft, layered stones are not very good as building stones. Other odd stones, such as the hard and flashy quartz, are hard to work with and rarely look natural.

Working with Mixed Lots

If you have a limited supply of stones, try mixing types to give added texture to a wall and keep it from visually fading into the landscape.

Finding a Source for Good Stone

Before you can work with stone, you have to get the stuff. And where you get it depends a lot on where you happen to be. However, whether you plan on buying stone from a stone lot or want to head out into the back woods of your property to collect loose rock, there are four points to remember that will simplify your task:

- Because one criterion for good stone is that it be as native to the environment as possible, start your search close to home.
- Especially when you're prospecting, look for usable shapes, such as a flat top and bottom, with the appearance you're looking for on what will be the stone's face. When you find some, there'll usually be more of the same nearby, because stone tends to fracture naturally along the same lines in a given area.
- If a stone looks doubtful for laying, pass it up. Bring home only the very best — you'll still have a lot of rejects. Every stone will fit somewhere, but not necessarily where and when you want it. It's not worth the added weight to haul a rock you won't use.
- If you're looking for a particular type of stone, seek the help of your state geology or mineral resource headquarters. Field geologists map stone underlayment and can tell you where certain stones occur.

Buying Stone from Commercial Vendors

Stone yards are an obvious place to begin. Most mine local stone for riprap (irregular stone used for fill) and crush it for roadway gravel. But they usually also import fieldstone and quarried stone for

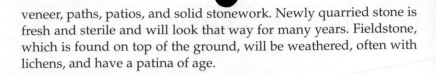

veneer, paths, patios, and solid stonework. Newly quarried stone is fresh and sterile and will look that way for many years. Fieldstone, which is found on top of the ground, will be weathered, often with lichens, and have a patina of age.

Prospecting for Stone in the Wilds

Fields and byways, woods and country roadsides provide plenty of rocks lying on top of the ground. That's where the quarries get their fieldstone. Go prospecting, but remember that the stone you find already belongs to someone; check with the landowners and get permission to remove it. You will need a pickup — even a flatbed truck if you will be transporting large stones — and one or two helpers, depending on the size of the stones.

It's cheaper to find your own stone than to buy it at the quarry. It's more work, too, but it gives you an excuse to get out into the country. So lay a sheet of plywood in the bed of your pickup truck to protect it, and have fun prospecting.

Please note: You absolutely must have permission from the local forest service or park management before removing stones from public land.

Finding and Recycling Secondhand Stone

Recycled stone from abandoned chimneys where houses have burned can be used. Sometimes these stones are free for the taking.

Most older structures were dry-laid, with just clay between the stones. The stones had to be good to stay in place, so there's little waste. Be sure to handpick; if you scoop them up with a loader bucket, you get a lot of dirt and rubble and other unusable stuff.

If you elect to dismantle a structure for the stone — a chimney or basement or house wall, for example — check first to see how it is put together. Lime-and-sand mortar, used for centuries until around 1900, can be dug out easily and allows the stones to separate nicely. Modern Portland cement, however, forms a strong bond, makes it hard to get the stones apart, and is difficult to remove from the stones themselves. If the stones are cemented with Portland, you'll have a lot of chiseling to do, and the stains will stay on the stones.

Rules of Etiquette for Private Lands

Excerpted from Natural Stonescapes, *by Richard L. Dubé and Frederick C. Campbell (Storey Publishing, 1999)*

There are many sources of stones on private lands. However, there are also some important standards of etiquette that you must be aware of before venturing out into the countryside on a stone-collecting mission:

1. Don't trespass. Ask the landowners for permission to explore the property.

2. Communicate with the landowners. Share with them what you are doing. The more they know, the more apt they are to let you onto their property.

3. Leave the site in better shape than you found it. For example, repair your ruts.

4. Never take more than you need.

5. Do not mark stone with spray paint; use a chalk or removeable marker, such as ribbon.

6. Be careful not to remove stones that could initiate or exacerbate an erosion problem.

7. Pay a fair price for the stones. This will vary by region and stone availability. You can find out what's a fair price by checking in with a local stone center, pit, or stone broker (a person who buys stone for redistribution to masons, stone centers, and landscape contractors).

When dismantling an old stone wall, to the greatest extent possible watch out for, and try to avoid disturbing, its inhabitants.

A chimney is usually startlingly easy to pull over, with a long cable and a tractor or four-wheel-drive vehicle. You get the chimney rocking and, once it's really off center, down it comes. Stones, dust, and mortar fly everywhere, so stay at least 100 feet (30 m) away.

A wall laid with lime mortar can be pried apart easily with a crowbar, one stone at a time. Sometimes sections can be pulled down with a tractor, but a lot gets lost in the rubble. Wear a dust mask around this kind of work, and never demolish stonework in a stiff wind.

Using Up Leftovers

Other sources may become available if you keep your eyes open and pursue leads. Many masons, for example, will gather stone from various places and sell to you in an off period. Naturally, they will want to keep the best for themselves, so insist on handpicking. Another source is leftovers from someone else's job. An offer to clean up the remaining stone from a building site may get you some valuable material.

Tools and Techniques for Handling Stone

Even in this high-tech age, most stoneworkers still pick up, load, stack, select, shape, and place stone by hand. In a single day, a mason may handle 10 tons (9,000 kg) of stone, one stone at a time, and many of the same stones several times.

Incorrect lifting

Correct lifting

Moving and Lifting

With a substance as unforgiving as stone, it is hard to overemphasize the importance of lifting properly. Learn to *squat and lift with your legs.* If you can't do it that way, don't do it at all. When in doubt, use heavy equipment to move and place stones.

For those stones you decide you can lift, the procedure is simple. Grab the stone in what would be a normal position, and then drop your rear another 2 feet (60 cm). When lifting, hug the stone close. It's a lot easier on your back and arms.

The wheelbarrow will be your handiest tool for moving small- to medium-size stones. You can lay it on its side, slide a stone in, and stand the wheelbarrow up by yourself to move it. (You may, however, need help laying the stone when you get it where you want it.) On rocky or steep ground, load the wheelbarrow back near the handles. You'll lift more, but the wheel will go over obstacles better. With the weight on the wheel, it's harder to push; even a pebble can stop you.

To load a large stone, lay the wheelbarrow on its side, slide the stone in, and pull the wheelbarrow upright.

You can also use simple plank "slides" to push heavy stones from the ground up to truck beds or the top of a wall. Just make sure the bottom of the plank is well anchored, set the top of the plank on the desired destination at a relatively low angle, and push

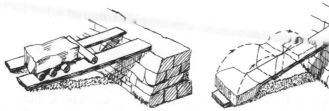

Large stones can be rolled or flipped end over end up a plank ramp to the top of your wall.

the stone up the plank, flipping it end over end. For large stones with a nice flat base, you can even place rollers between the stone and the plank, which will allow you to simply slide the stone up the plank.

Moving the Big Ones

For really big stones (those requiring more than two people to handle), you may want to use machinery. This includes everything from a single pry bar, block and tackle, or come-along (a hand-operated ratchet hoist) to an electric winch, a hydraulically activated boom mounted on a truck, a hydraulic jack, or a tractor with bucket. It depends on where you are and how much room you have to maneuver. Unlike big commercial construction sites with flat, uncluttered access, most stone-gathering and stone-laying sites are among trees, on sloping ground, or against walls, where there is little room.

Lifting and loading devices for large stones are many and ingenious. You will seldom have heavy equipment out in the woods where you find stone, but you can rig a tripod, even if there are no trees handy. Make it tall, of doubled 2-by-4s about 12 feet (3.7 m) long. Wrap a chain around the apex and hold it in place with some 20-penny spikes. Hook a come-along into this chain and wrap another chain around the stone. Set the legs of the tripod far enough apart to allow your pickup truck to back between. Lift the stone, back under with the truck, and let the stone down. The process isn't fast, but a large stone will cover more area than several smaller ones and will look better, too.

Of Trucks and Loads

A two-wheel-drive, six-cylinder pickup truck is an ideal all-around construction vehicle; it can carry 1½ tons (1,362 kg) of stone — 2 tons (1,816 kg) on short hauls. A heavier truck will use more fuel; a lighter one won't haul as much.

Don't use an automatic transmission if you plan to haul stone in hilly country. It's not much of a problem going uphill, but coming down is scary. The lowest gear just won't hold back a pickup loaded with stone on steep roads.

One of the simplest loading devices is a small swiveling crane that mounts in a corner of a pickup bed. It is raised by a simple hydraulic jack and will lift more weight than you should subject the average pickup to. You can block under the rear bumper and load large stones with this little crane, but watch the limits of mounting bolts and braces.

You don't *have* to use big stones at all, but the stonemason's adage that a small one takes just as long to lay as a big one is largely true. And big stones look good. Generally, your work will be just fine with maximum two-person rocks, which can easily be 2-foot (60 cm) expanses and maybe 6 inches (15 cm) thick.

As you work with various stones, you'll come up with your own ingenious devices for getting them into place. In general, you can move small quantities of stone by hand in the time it takes to set up machinery. If you're in reasonable physical shape and don't overdo it, you can get by with a wheelbarrow and pickup truck. If you do a lot of work, consider the advantages of more equipment. Stonemasons are a lot like anglers and mechanics, though — they collect all sorts of tools they rarely use. Unless you plan to do a great deal of stonework, don't bother investing in expensive machinery.

Staying Safe

Safety equipment is essential. Use a hard hat, goggles, gloves, steel-toed boots, and a dust mask around stone. Sand, chips, and dust are always around. And no matter how careful you are, use goggles; a stone chip will eventually fly straight at your face.

Gloves help keep your hands from getting scratched, but not your fingers from getting smashed. (Your toes will be more fortunate if you encase them in steel-toed boots.) Your hands do get rough without gloves, and in winter tend to crack, chap, and bleed no matter how much lotion you use.

Safety goggles and sturdy work gloves are essential for stonework.

Cutting and Shaping Stone

Good fieldstone masons will only true up — that is, make level or plumb — the shapes they find, not try to create new ones. A long, triangle-faced stone might get its corners chipped to match others in a wall. Or a curved edge could get straightened for a better fit in a ledge pattern. A thick stone might need to be split for veneer work — and if you're lucky, you'll get two stones. Minimal shaping is always recommended, because newly cut stone faces glare from an aged wall. It takes years for these fresh cuts to blend in, so try to avoid them where they will show.

Using a Hammer and Chisel

To cut a corner off a stone, lay the stone on something soft to absorb the shock. Sand in a box is good, or use a table padded with old rugs. Try to get the work area elevated to about counter height (36 inches; 90 cm). Bending down to work at ground level is hard.

To mark the cut you want, make a series of light hits with a striking hammer on a stone chisel, then go over this line again, this time striking harder. After about the third pass, turn the stone over and mark that side, too. Repeat the procedure, using more force with each repetition. If you're working near the edge of the stone, lean the chisel out a bit to direct it into the mass of the stone. (If you hold it perpendicular, the stone will tend to chip off instead of breaking all the way through.)

Remember that a light tap will tend to chip the stone out toward the edge; a heavier hit will crack it deeper, usually closer to where you want it. Sometimes the stone will break toward the edge anyway, leaving a ridge on the edge face. Slope the chisel sharply, just enough for it to bite into the surface, striking into the mass near the ridge. Take off small chips from both sides this way until you have dressed the surfaces sufficiently. Ideally, the stone will break all the way through from marked line to line, but don't count on it the first few thousand times.

A large stone-breaking maul, which has an edge to it, will create small stones from big ones. Use a 12-pounder (5.4 kg) for this if there's no way to utilize big stones as they are. Hit the line lightly, as with the chisel, then go back over it harder. You'll have to smooth your breaks afterward with the hammer and chisel, because the

Keep Your Equipment in Good Condition

Watch the condition of stone chisels, particularly the struck ends. These will mushroom from repeated strikes with the hammer. Grind the ragged edges off your chisels as you notice them forming. If the chisel head does *not* mushroom in use, be wary: It may be too hard to spread and therefore hard enough to shatter. And remember, always wear goggles when using or grinding stone chisels.

maul gives you approximates only. Splitting sedimentary stones, such as sandstone and limestone, is quick with a maul if your aim is good. It's a lot like splitting wood, but it doesn't go quite as quickly.

The Elusive "Perfect Fit"

Masons have always been fascinated by cutting mortises and steps in stones for the perfect fit. It's rarely worth the time it takes, and too often a stone breaks after hours of shaping. Try to use the spaces between stones creatively rather than going for a tight fit every time. Often, you can proceed much more efficiently if you lay a stone that leaves a space you can fill with a shim or chip.

The dry-stack look, which is popular right now, requires closer fitting and therefore more shaping. Dry-stack is mortared stone, but none of the mortared joints shows, so it looks like tightly laid dry-stone work. Sometimes it will allow more chips and shims, sometimes not. If large stones are required for a job, there are just two solutions: Search more or shape more.

For a tight fit at the edges, you may have to take down a hump. To do this, stand the stone on edge and dress the offending surface with a chisel or a stone point, which is just what it sounds like: a chisel that comes to a point. It is used to chip off bits of stone. Slow, but it works.

The most important thing to remember about shaping stones: Don't attempt it unless you must. If you have a lot of extra stone, you may be able to find the right fit, even if you must use two or more to fill the gap. You'll find cutting frustrating enough to want to avoid it, even with easily shaped sandstone. If you're using harder stones, like granites, just find the right ones.

Building a Drystone Wall

A freestanding drystone wall is the simplest and most attractive structure you can build of stone. There's no footing, no mortar, no cracking with freezing. If you use stones gathered from the top of the ground, they'll have lichens and an aged appearance. So a new drystone wall will look as if it's centuries old.

Building a basic drystone wall means simply placing stones on top of other stones, with some intelligent constraints. (If it falls down, for instance, it isn't a wall.) A freestanding drystone wall should have stones on each face that slope inward against each other. Each stone, if possible, should push against its neighbor in a controlled situation that doesn't let either move. Thus, the stone in drystone walls stays in place without mortar because gravity and friction hold it in place — if it has been laid properly.

Materials and Tools

To build a freestanding wall, 3 feet (90 cm) high and 2 feet (60 cm) thick, you'll need the following:

Materials

1 ton (0.9 t) of relatively flat (tops and bottoms) stones, 6–24 inches (15–60 cm) wide and 2–6 inches (5–15 cm) thick, for every 3 feet (90 cm) of wall length

Tools

Stone chisel	Tape
Striking hammer	4-foot (1.2 m) level
Mason's hammer	Pry bar (straight or crowbar)

Stone by Weight and Volume

One ton (0.9 t) of stone makes about 3 running feet (90 cm) of stone wall, 3 feet (90 cm) high and 2 feet (60 cm) thick. You can buy a ton of stone at a stone and gravel yard or some home and garden centers. If you gather stone in the field, a ton of stone is about 17.5 cubic feet (3.5 by 5 by 1 feet, or 1.1 by 1.5 by 0.3 m) or a full-size pickup truck loaded about 6 inches (15 cm) deep.

Step 1: Dig a Trench for the Base

The stability of a drystone wall is based on gravity; the stones lean into each other, thereby holding each other in place. How do you get the stones in your wall to slope inward? Dig the topsoil at the base of the wall into a shallow trench that's 24 inches (60 cm) wide and 4 to 6 inches (10 to 15 cm) deep, but with a slight V-slope; it should thus be about 2 inches (5 cm) deeper in the center. Keep the ditch level lengthwise; if the ground slopes, step the trench to keep it level.

The trench should form a slight V-shape.

If you're building a wall on sloping ground, the trench must be stepped, or built like steps, so that each stone in the first course (and remaining courses) can be set level.

Step 2: Lay the First Course of Stones

Walls are built in courses, or layers. To lay the first one, place stones along the bottom of the trench in pairs, with each sloping toward the center. Use stones that have a relatively even outside edge and that reach approximately to the center of the trench. If a stone extends 2 to 3 inches (5 to 8 cm) past the center, adjust the soil for it, and use a narrower one opposite it. If both stones are short, fill the center space with broken stones.

Place uneven surfaces down on this first course, digging out as necessary. Leave as smooth a top surface as possible, matching stone heights.

Leave as smooth a surface as possible on the first course.

Calculating the Appropriate Thickness

The ratio of thickness to height in your wall depends on how "good" the stones are. A good stone is one that is level on top and bottom so that it will lie flat and support the maximum weight. With good tops and bottoms, stones in a wall will stay put with a thickness that is half the height of the wall. So a 4-foot wall, for example, can be 2 feet thick.

Step 3: Lay the Remaining Courses

Begin the next course with similar stones, taking care to cover the cracks between the first-layer stones. If you have spanned the 24-inch (60 cm) trench with one 15-inch (38 cm) stone on, say, the right side and a 9-inch (23 cm) one on the left, reverse this now. If the stones were 12 inches (30 cm) long (along the length of the wall), use shorter or larger ones on this second layer to avoid vertical running joints. Try to use stones of uniform thickness (height) in each course. Where this is impossible, use two thin ones alongside a thick one for an even height.

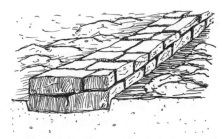

Span cracks in the first course with solid stone in the second course to avoid vertical running joints.

Place a 24-inch (60 cm) stone as a tie-stone (see the box at right) across the width of the wall every 4 feet (1.2 m) along a course and in every third course in any given location in the wall. Because this will be seen on both wall faces, it should have relatively straight ends. Use the hammer and chisel, if necessary, to shape these faces.

When you use a tie-stone, it probably won't have a convenient dip in the middle, so you'll have to reestablish the V-slope with the next course. Tapered stones are the obvious answer, but you can also wedge up the outside edges with chips, or shims. These shims must not themselves be tapered, however, or they will gradually work their way out of the wall with the flexing and movement inherent to drystone work. Use thin rectangular chips from your shapings or break thin stones for shims. Properly laid, wedged-in shims and stones won't shift, even if they are used as steps, where they get jostled a lot.

Wedges help reestablish the sloping angle that holds the stones in place.

Working with Tie-Stones

In the cross section of a freestanding wall, you see two walls leaning into each other. However, because this is a precarious balance in which the two walls can come apart, masons tie across the wall whenever possible with a *tie-stone,* a stone that spans the width of the wall and provides stability and support.

Tie-stones must be laid completely flat, or they will creep downhill with freezes and thaws. They're especially important for capping the wall (in which case they're called capstones), for they hold it together and keep out rain, which will freeze and push the wall apart. Capstones also make it more difficult for dust, leaves, and other debris to blow into the wall, where they can nurture tree seeds; allowing tree roots to sprout in your wall is guaranteed to force it apart.

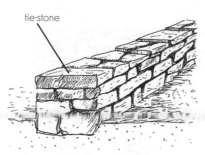

Tie-stones placed along the courses span the width of the wall.

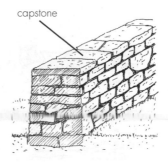

Capstones are tie-stones placed on top of the wall.

Step 4: Cap the Top

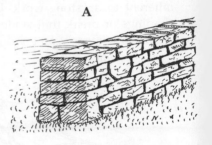

Begin and end the wall vertically (A), or step it down to the ground (B). If the ground rises, keep the wall top level until it fades into the grade (C). Use as many tie-stones as possible for the top layer. These are called capstones, and they are easily dislodged unless large and heavy. The best stones should be saved for capstones, because piecing this top layer will make it unstable.

Five Secrets for Success with a Drystone Wall

1. Establish an even plane. The surface of a stone wall is never even, but if you establish a plane that you go back to often, the stones in the wall can jut out or be recessed without appearing sloppy. If you've been careful to align the outer edges of the stones in the wall, the tie-stone ends can vary in appearance and overhang without seeming out of place.

2. Use step patterns on sloping ground. When building a drystone wall that runs downhill, keep the stones level by removing the topsoil at the base of the wall in a stepped pattern (see page 15) and repeating this pattern with the capstones. Stones that slope or are laid at an angle will move over time, and even though it may take them awhile to do this, you don't want your wall to fall down. Build it right the first time.

When working with a stepped base (see page 15), the top of the wall must be stepped as well.

B

C

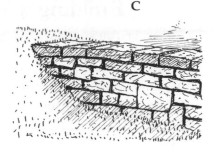

3. Lay stones with their best edges out. Lay the slightly sloped stones with their best edges out. If that leaves gaps in the center of the wall, fill these with rubble — that is, any scrap stone not usable otherwise. More than likely, your supply of neat stones with nice edges will be limited or nonexistent, so you'll have to shape at least one end for a face (see page 12).

4. Work with thin stones. It's simpler to build a drystone wall with thin stones; they're easier to handle and can be shaped more easily than thick ones.

5. Avoid vertical running joints. Always lay stones side by side with the top edge parallel to the ground, and make sure to cover the crack between them with a stone on the next course. If you don't, you'll have a running joint, or vertical crack, which will weaken the wall too much. A joint that is two courses deep is permissible but not desirable.

Vertical running joints will compromise the strength and durability of your wall and should be avoided at all costs.

Building a Mortared Wall

A mortared wall seals out water and roots, does not flex appreciably with temperature changes, and is much stronger than a drystone wall. Properly laid on a footing and with joints raked ½ to 1 inch (13–25 mm) deep to let the stones stand out, a mortared wall is very attractive. It's more than a dry wall with cement in it, however.

Materials and Tools

To build a freestanding mortared wall, 3 feet (90 cm) high and 1 feet (30 cm) thick, you'll need the following:

Materials

1 ton (0.9 t) of stone up to 12 inches (30 cm) wide, of varying thickness and length, with as many square, flat surfaces as possible, for every 6 feet (1.8 m) of wall

Concrete: About 2½ cubic feet (71 cu dm) of ready-mix concrete for every 1 foot (30 cm) of wall length, or about 1 cubic yard (0.8 cu m) for every 10 feet (3 m) of wall. If you choose to mix the concrete yourself, you'll need about 1 cubic yard (0.8 cu m) of gravel, ⅔ cubic yard (0.5 cu m) of sand, and 9 sacks of Portland cement for every 10 feet (3 m) of wall.

Mortar: 3 sacks of Portland cement, 1 sack of mason's lime, and 1 ton (0.9 t) of sand for every 12 to 15 feet (3.7 to 4.6 m) of wall length; water

If wall runs up- or downhill, 8-by-30-by-1-inch (21 by 76 by 2.5 cm) boards for bulkhead

½-inch (13 mm) reinforcing rod, sold in 20-foot (6 m) lengths; to be set in pairs into the footing running the full length of the wall

Rebar grade stakes, 1 for each 4 feet (1.2 m) of wall length

Tools

Two 48-inch (1.2 m) stakes (one at each end for layout string to maintain a straight wall)	⅜-inch (1 cm) pointing tool
	Wire brushes
	Stone chisel
String	Striking hammer
Wheelbarrow	Mason's hammer
Pick	Pry bar
Shovel	4-foot (1.2 m) level
Hoe	Tape
Large trowel	

Because it cannot flex, mortared stonework must be set on a footing, or base, that extends below the frost line. The footing can be as high as the surface of the ground or as low as 6 inches (15 cm) below it. A footing, which is wider than the wall, distributes the wall's weight over a larger area, reducing its downward pressure. Building codes vary, but usually anything that calls for a foundation such as a footing requires a building permit.

A 3-foot-high (90 cm) mortared stone wall can be narrower than a comparable drystone wall; a width of 12 inches (30 cm) provides adequate stability for a straight wall. The footing should be twice the width of the wall thickness.

Storing Cement Ingredients

Because of the time it takes to mix and pour, and the resulting small batches of concrete that you work with, you won't be able to mix and pour all of the cement required by a wall at once, so be sure to store Portland up off the ground, wrapped in plastic or under a roof. Don't secure the plastic to the ground or ground moisture will condense under it and get the cement wet. And don't store Portland for more than a month. There's often enough moisture in the air to start the chemical process of setting up, which you don't want until after you mix and pour.

Step 1: Prepare a Trench for the Concrete Footing

To lay a footing, you'll first need to dig a ditch 24 inches (60 cm) wide to below the frost line in your area. Building inspectors require that you set rebar grade stakes in the ditch to determine the thickness of the footing. Drive these in every 4 feet (1.2 m) or so, and level their top ends at the height to which you will fill the ditch with concrete. Use a 4-foot level, line level, water level, or transit for this. The inspector will want to see these set, along with stop bullheads and smooth, solid ditch bottoms and sharp ditch corners, before he or she will approve the pour.

Laying Gravel Footings

Another type of footing that is much cheaper, but not as long lived or as strong, is a gravel footing. With this, you dig a ditch well below frost line but only as wide as your wall thickness. Then you dump in gravel or crushed stone to a minimum of 6 inches (15 cm) deep — but you must keep the top below frost line. Then level off and start laying stone.

The theory here is that water won't get into the wall and freeze, because the mortar seals it out. It doesn't freeze below the wall and cause it to buckle because it drains to below the frost line. And because the water soaks through the gravel and runs off quickly, the wall stays stable.

There are, however, two drawbacks to this kind of footing. It does nothing to distribute the weight of the wall, which can settle and crack. And eventually the gravel gets dirt washed into it, the water stops percolating through it, and the wall essentially sits on dirt. In addition, tree roots do a lot more damage when they can get under a wall this easily. But a gravel footing can serve you well for a long time, and given the expense of the concrete footing, a gravel-based wall may be the way to go.

Step 2: Calculate How Deep the Footing Should Be

The depth or thickness of the footing concrete is largely up to you, beyond the 6-inch (15 cm) minimum for a hypothetical 3-foot-high (90 cm) wall. It is commonplace to lay concrete blocks below ground level on the footing instead of wasting good stones where they will not be seen. In our situation, it's actually cheaper to buy ready-mixed concrete and fill the ditch to ground level. Additional labor and materials to bring the wall up to the surface are too expensive.

If you're in a place a concrete truck can't access and you're mixing concrete for the footing by hand or in a mixer, however, the equation can change, especially if your wall is going to be a long one, in which case it may take too much concrete to fill the ditch that way. If you have lots of good stone, you can afford to hide some of the less-than-gorgeous pieces belowground. Keep in mind, though, that really misshapen stones hidden down there won't hold up the wall properly unless they're embedded in concrete.

The quantities given in the materials list on page 20 assume you will dig to 18 inches (45 cm) and fill to the grade. Of course, you can substitute 12-inch (30 cm) concrete blocks or stone below grade, on the footing. However, as noted on the previous page, it is quicker and not too much more expensive to fill the trench to grade.

Step 3: Mix the Concrete

You can buy ready-mixed concrete at most building supply and garden shops, but it tends to be slightly more expensive than mixing your own. Cost versus convenience — it's your choice. If you're mixing your own, the basic mix is:

1 **part Portland cement**
2 **parts sand**
Water
3 **parts gravel, 1 inch (2.5 cm) or less in size (what quarries call six-to-eights)**

Calculate how much footing you'll need before you start. At 2 feet (60 cm) wide and a minimum of 6 inches (15 cm) deep, you'll be mixing and pouring at least 1 cubic foot (28 cudm) per running foot of ditch. For a 50-foot (15 m) wall, that's a minimum of 50 cubic feet (1.4 cu m) of concrete (more, if it has steps in it), for which you'll need about 2 cubic yards (1.5 cu m) of gravel, allowing for spills and settling. That amounts to several loads in your pickup truck, so you might want to have the quarry deliver it. Its crushed rock generally makes stronger concrete than rounded creek gravel, too, although you can dig the latter yourself. Add to that two thirds as much sand and about 13 cubic-foot sacks of Portland.

Mix concrete footings in small quantities so that the concrete won't dry out faster than you can use it; if you have a wheelbarrow, you'll be mixing 1 cubic foot (28 cu dm) at a time; if you have a small-to medium-size mixer, you'll be mixing 2 to 3 cubic feet (56–84 cu dm) at a time. Start with 4 shovels of sand, then add 2 shovels of Portland cement. Dry-mix, then add water until the mix is wet and loose. How much water you need varies a lot; the key factor is how dry or wet the sand is when you start. Add roughly 6 shovels of gravel last, working it into the wet mix a little at a time.

The finished concrete should be dry enough to hold shaped peaks when you shovel it into place, but wet enough to level out when it is shaken with a hoe. If water puddles up on the mixture, it is too wet. Excess water will leave air pockets when the concrete dries, which will weaken it.

The concrete mixture should be firm enough to hold stiff peaks.

Step 4: Pouring and Reinforcing the Concrete

You'll need reinforcing rods, or rebar, typically ½ inch (13 mm) thick, in the footing to strengthen it. Steel makes cement strong. Otherwise, soft spots in the ditch bottom, or places where it goes from bedrock to dirt, can settle and crack your stone wall. Set the rebar in the concrete halfway up the footing thickness. Don't prop the rod up on bricks or rocks, because that leaves a hairline crack for water to seep into, which will rust the steel. Instead, pour 4 inches (10 cm) of concrete, then place two rods 1 foot (30 cm) apart in the center of your 3-foot (90 cm) ditch. Overlap the rebar ends 6 inches (15 cm). Then pour the rest of the footing.

Quick Tip

Always try to pour the concrete all at once. If a section has as much as an hour to set up before you join it with another, you get a "cold" joint, which is another place for water to seep in and thus weaken the concrete.

Establishing a rough level is close enough if you're laying stone on top of the footing, but it must be smoother for concrete blocks. You can dump wheelbarrow loads in, then smooth them with a hoe. As you even out the surface, feel around for the tops of the rebar grade stakes that mark off a level surface.

Let the cement dry for 2 days.

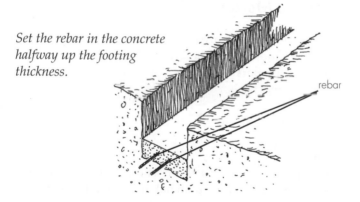

Set the rebar in the concrete halfway up the footing thickness.

rebar

Step 5: Plan the Layout

After 2 days, remove bulkhead boards, if any. Whether you've built to ground level with concrete, concrete blocks, or stone, you're ready now to lay stone that will show. Your work will be judged by this, so it needs to be right. For strength, a basically horizontal or ledge pattern is best. Too many same-size stones look monotonous, so vary the sizes often. Just remember to return as much as is practical to the horizontal arrangement.

Dry-fit 3 to 4 feet (0.9–1.2 m) of wall length, keeping the outside faces even and top surfaces level. Be sure each stone will stay in place dry. Shape with hammer and chisel where necessary. Leave a ½ inch (13 mm) space between each stone. You may use two stones for the 12-inch (30 cm) wall thickness.

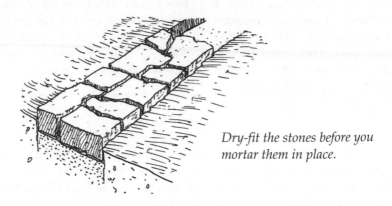

Dry-fit the stones before you mortar them in place.

Step 6: Mix the Mortar

A heavy contractor's wheelbarrow is ideal for mixing mortar. You can move it to the sandpile, to where your Portland cement and lime are stored, and to the part of the site you're working on. The basic mix for stonework mortar is:

9   **parts sand**
2   **parts Portland cement**
1   **part lime**
     Water

Start with the sand and add the Portland cement and lime, so that the wheelbarrow is no more than half full. A medium-size shovel used for the measuring produces a good batch of mortar.

Mix dry first, using the shovel, a hoe, or both. Then move the dry mix away from one end of the wheelbarrow pan and pour in about ½ gallon (1.9 l) of water.

With the hoe, begin "chopping off" thin slices of the dry mix into the water. Work each bit until it is wet throughout, then chop some more. When this water is used up, open a hole beyond the mixed mortar for more water and repeat the process. Be sure you don't leave dry pockets of mix down in the corners as you go. When you get to the end of the wheelbarrow using this process, you should be finished. Take care not to add too much water at a time or the mix can get overly wet. It should stand in peaks.

If the sand is wet from rain, you'll need very little water for the mix. As you get near the end of the batch, use only a small amount of water at a time. It's easy to go from too dry to too wet with just a cupful of water.

Mix mortar by "chopping off" slices of dry mixture and working each slice with water.

The consistency should be as wet as possible without running or dripping. If the mix is too wet, add sand, Portland cement, and lime in proportionate amounts and mix until the batch stiffens up. Or you can leave it for 20 minutes or so; excess water will float to the surface, where you can pour it off. The mortar at the bottom will thicken enough for use, and you can scrape aside the soft stuff to get to it.

As you use the mortar, it will dry out more. This will help amend too-wet mortar and will mean you have to add water to an ideal mix. Use all mortar within 2 hours of mixing.

Step 7: Mortar the Stones in Place

Follow the principles of drystone work: Minimize mortar but seal out water and make the wall rigid. If you elect to try for a dry-stack, mortarless look, you have a big job ahead of you. Aim for joints between ½ and 1 inch (13–25 mm), recessed deeply (1 inch or so) so that they're not noticeable. A substantial mortar joint will bond well, keep out water better, and take up the irregularities in the stones more easily.

Lay a base of ¾ inch (2 cm) of mortar on the concrete footing, removing and replacing the prefitted stones as you go. Rock each stone just a bit to work out air pockets. If mortar gets pushed out to the face, trim it off and rake out the joint with the pointing tool. Don't let mortar get onto the stone faces at all; the stains are hard to get off. Fill between the stones, using the pointing tool to push mortar off the trowel into the cracks. Recess visible joints at least ½ inch (13 mm). You may extend the first course as far as you like before beginning the second one. Keep the mortar wet for at least 2 days.

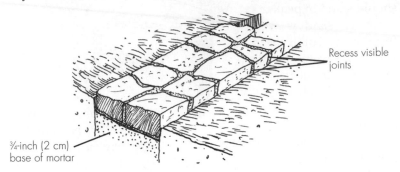

Recess visible joints of mortar at least ½ inch (13 mm).

Repeat the dry-stacking and subsequent mortaring of stones for the entire wall, mixing new mortar as necessary. Step hill walls and lay stones flat instead of up on edge unless they are very thick. Lay several feet horizontally, working with the full width of the wall so that you create the outside faces together. Then go back and work on the second course.

For the second and succeeding courses, avoid exterior and interior vertical running joints, just as in drystone work. Here, too, tie-stones are necessary. For this narrow wall, there will probably be no need to fill the center with small pieces. Fill small gaps with mortar, for strength. Work out all air spaces by packing with the pointing tool.

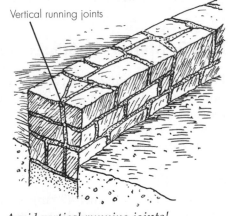

Avoid vertical running joints!

Step 8: Rake the Joints

Within 4 hours of applying mortar, use the pointing tool to rake the mortar out the joints onto a trowel for reuse. The raked-out joints should have a uniform depth of at least 1 inch (2.5 cm); no matter how deeply you want the mortar joints struck, or recessed, this depth should be consistent. It's probably impossible to use just the right quantity of mortar every time, so count on raking some out every day. At this point the mortar will probably still be wet enough that it will smear, so don't expect to do a complete job now.

Within 4 hours of applying mortar, use a pointing tool to rake out the joints to a consistent depth.

Keeping the Wall Clean

Don't let mortar get onto stone faces, if at all possible, because it can be hard to clean off. It's a bit tricky to keep mortar dry enough that it won't run, however, because mortar has to be wet enough to bond to the stone surfaces. Stone laid up with very dry mortar comes apart. The mortar should be stiff enough that peaks don't shake down flat unless vibrated very hard.

If you get a run, use a sponge while the mortar is still wet. Dried stains can eventually be removed by scrubbing with water and a wire brush.

For greater ease in removing the dried stains from your wall, you can add muriatic acid to the water (1 part muriatic acid to 10 parts water). Repeat the acid treatment until the stain is gone. Flush with lots of water about 30 minutes after each application.

Step 9: Clean the Joints and Faces

After the mortar is dry enough not to smear, use a wire brush to clean up and make the mortar neat. Raking will leave grooves, pits, and uneven places, but the wire brush will smooth them. Wet the mortar thoroughly afterward, using any method you like, when you're sure that the mortar is dry enough not to run down the faces of the stones.

A wire brush and water will eventually remove mortar stains from stone.

Use the wire brush for additional cleanup. Keep the mortar wet with a spray, or slosh water on, for at least 2 days. Normally four or five wettings a day will suffice.

---

7 Secrets for Success with Mortar

1. Lock the ends. If your wall has a freestanding vertical end — if it doesn't end against a building or fade into the ground — alternate short and long stones so that the end is even but each course is locked together. It's like using half bricks to end every other course. Cover the crack between two stones with the stone above.

2. Keep the wall straight. Most masons use stretched string to keep the wall straight.

You can use a guide string to keep your wall straight. Just tie the string to tall stakes driven in the ground, and elevate the string as necessary.

3. Test before mortaring. Dry-fit a few stones to see how they look before you actually mortar them. Don't do too many, though, for the

Step 10: Cap the Wall

It isn't necessary for all the capstones to be large and heavy, because they are mortared in place, but make sure mortared joints are tight, so that water won't get into the wall and freeze. If it does, stones will break off the wall.

A mortared hill wall need not be stepped on top. The individual courses should be level, and if odd-shaped accent

In contrast to a drystone wall, a mortared wall built on a stepped trench on sloping ground does not need to be stepped on top.

stones are used, return to a level again. If the top is to be sloped instead of stepped, you must use sloped stones — those thicker on one edge — and lay them flat on the stones below them.

---

mortar will change the dimensions. Then set this half dozen or so, rake the excess mortar from the joints, and stand back and admire your handiwork.

4. Fill wide joints with stone chips. Fill any wide mortar joints with appropriately shaped chips. There will be a lot of bits from your shaping, but few of them will be shaped to the places they will fill. So lay an almost-good stone, and count on a shim to take up the slack.

5. Avoid vertical running joints. After laying stones side by side on one course, be sure to cover the crack between them with a stone on the next course. If you don't, you'll have a running joint, or vertical crack, which will weaken the wall too much. A joint that is two courses deep is permissible but not desirable.

6. Prepare for the next day's work. When you stop for the day, leave stone in steps, rather than leaving a vertical end, so it's easier to tie onto the next time.

7. Keep the mortar wet. Moisture is important to curing mortar, which is a chemical process that goes on for several days. If the mortar dries out too soon, the process stops and the stuff has little strength. So wet it thoroughly about four times during the day after you lay mortared stone. In very hot weather, drape plastic sheeting over it to hold in the moisture. After 2 days, it doesn't matter much; the process naturally slows down.

Other Storey Titles You Will Enjoy

Garden Stone, by Barbara Pleasant.
A best-selling guide to 40 enchanting, creative
projects for landscaping with plants and stone.
240 pages. Paper. ISBN 978-1-58017-544-9.

How to Build Paths, Steps & Footbridges,
by Peter Jeswald.
The fundamentals of planning, designing, and constructing
creative walkways for your home landscape.
248 pages. Paper. ISBN 978-1-58017-487-9.

Natural Stonescapes, by Richard L. Dube &
Frederick C. Campbell.
More than 20 designs for stone groupings appropriate
for all types of landscapes, from flat lawns to hillsides.
176 pages. Paper. ISBN 978-1-58017-092-5.

Outdoor Stonework, by Alan and Gill Bridgewater.
A definitive guide to working with stone, from selecting
and cutting stone to preparing foundations
and completing projects for your yard.
128 pages. Paper. ISBN 978-1-58017-333-9.

Stone Primer, by Charles McRaven.
The essential guide for homeowners who want
to add the elegance of stone, inside and out.
272 pages. Paper. ISBN 978-1-58017-670-5.

Stonework: Techniques and Projects,
by Charles McRaven.
A collection of 22 fully illustrated, step-by-step instructions
to teach the tricks of the trade from a master craftsman.
192 pages. Paper. ISBN 978-0-88266-976-2.

These and other books from Storey Publishing are available
wherever quality books are sold or by calling 1-800-441-5700.
Visit us at *www.storey.com* or sign up for our newsletter
at *www.storey.com/signup.*